海底地名命名标准

B-6 出版物　4.2.0 版

英文/中文版

STANDARDIZATION OF UNDERSEA FEATURE NAMES

Publication B-6

Edition 4.2.0

国际海道测量组织　编
政府间海洋学委员会

国家海洋信息中心　译

中国海洋大学出版社
·青岛·

图书在版编目（CIP）数据

海底地名命名标准：B-6出版物4.2.0版：英文、汉文 / 国际海道测量组织，政府间海洋学委员会编；国家海洋信息中心译. —青岛：中国海洋大学出版社，2024.6
书名原文：Standardization of Undersea Feature Names:Publication B-6 edition 4.2.0
ISBN 978-7-5670-3823-3

Ⅰ.①海… Ⅱ.①国… ②政… ③国… Ⅲ.①海底—地名—命名—国际标准—英、汉 Ⅳ.①P72-65
中国国家版本馆CIP数据核字（2024）第067221号

©Copyright International Hydrographic Organization (2018)

This work is copyright. Apart from any use permitted in accordance with the *Berne Convention for the Protection of Literary and Artistic Works* (1886), and except in the circumstances described below, no part may be translated, reproduced by any process, adapted, communicated or commercially exploited without prior written permission from the International Hydrographic Organization (IHO). Copyright in some of the material in this publication may be owned by another party and permission for the translation and/or reproduction of that material must be obtained from the owner.

This document or partial material from this document may be translated, reproduced or distributed for general information, on no more than a cost recovery basis. Copies may not be sold or distributed for profit or gain without prior written agreement of the IHO and any other copyright holders.

In the event that this document or partial material from this document is reproduced, translated or distributed under the terms described above, the following statements are to be included:

"Material from IHO publication [reference to extract: Title, Edition] is reproduced with the permission of the International Hydrographic Organization (IHO) (Permission No…/…), which does not accept responsibility for the correctness of the material as reproduced: in case of doubt, the IHO's authentic text shall prevail. The incorporation of material sourced from IHO shall not be construed as constituting an endorsement by IHO of this product."

"This [document/publication] is a translation of IHO [document/publication] [name]. The IHO has not checked this translation and therefore takes no responsibility for its accuracy. In case of doubt the source version of [name] in [language] should be consulted."

The IHO Logo or other identifiers shall not be used in any derived product without prior written permission from the IHO.

本出版物版权所有。除依据《伯尔尼保护文学和艺术作品公约》（1886年）许可的使用和下述情形外，未经国际海道测量组织（IHO）事先书面许可，本出版物的任何部分均不得用于翻译、任何形式的翻印、改编、传播或商业性使用。本出版物中的某些内容可能归属于其他组织，翻译和/或翻印必须获得该所有者的许可。

本出版物的全部或部分内容可翻译、翻印或作为一般参考资料发行，但售价不可高于成本。未经国际海道测量组织（IHO）和其他版权所有者事先书面同意，不得通过出售或发行该出版物的副本获取利润。

如依据上述条款翻印、翻译或发行本出版物的全部或部分内容，应包含以下声明：

"经国际海道测量组织（IHO）许可（许可号……/……），本材料翻译于国际海道测量组织（IHO）出版物［节录的书目：标题、版本］，但该许可不对翻印材料的正确性承担责任：如有疑义，以国际海道测量组织（IHO）发布的权威文本为准。对源于国际海道测量组织（IHO）出版物材料的汇编不构成国际海道测量组织（IHO）对本产品的认可。"

"本［文件/出版物］是国际海道测量组织（IHO）［文件/出版物］［名称］的翻译版，国际海道测量组织（IHO）并未对本翻译内容进行检查，因此不对其准确性承担责任。如有疑义，请查阅［名称］的［语言］原文。"

未经国际海道测量组织（IHO）的事先书面许可，不得在任何衍生产品中使用国际海道测量组织（IHO）徽标或其他标识。

出版发行	中国海洋大学出版社	社　　址	青岛市香港东路23号
邮政编码	266071	网　　址	http://pub.ouc.edu.cn
出 版 人	刘文菁	责任编辑	张　华
电　　话	0532-85902342	印　　制	青岛国彩印刷股份有限公司
版　　次	2024年6月第1版	印　　次	2024年6月第1次印刷
成品尺寸	185 mm × 260 mm	印　　张	3.25
字　　数	65千	印　　数	1~1000
定　　价	48.00元	订购电话	0532-82032573（传真）

发现印装质量问题，请致电0532-58700166，由印刷厂负责调换。

《海底地名命名标准》

李艳雯　邢　喆　郭灿文　**译**

毛　彬　李四海　**审译**

译者的话

海底地名分委会（SCUFN）是由国际海道测量组织（IHO）—政府间海洋学委员会（IOC）下设的全球通用大洋水深制图计划（GEBCO）联合指导委员会（GGC）任命的专业组织。该组织成立于1993年，主要关注全部或者主体（50%以上）位于领海以外的海底地理实体命名，是当今海底地名领域具有较高权威性和国际影响力的国际组织。海底地名分委会（SCUFN）每年召开一次会议，主要审议各国提交的海底地名提案。提案一经审议通过，将被收录至GEBCO海底地名词典中，应用于全球大洋水深制图中。

为了有效地组织指导海底地名工作，海底地名分委会（SCUFN）于2008年由国际海道测量局（IHB）发布了《海底地名命名标准》（B-6出版物4.0.0）版本。此后历经修改，分别于2013年9月重新发布了该标准的4.1.0版本，并于2019年10月发布了4.2.0版本。截至目前，该标准已先后发布了英文/法文、英文/西班牙文、英文/韩文、英文/日文、英文/俄文、英文/葡萄牙文和英文/中文等版本。

国家海洋信息中心作为我国海底地名技术支撑单位，业务化开展国际海底地名态势跟踪、地名提案编制审核、海底地名网运维等工作。为规范我国海底地名提案编制工作，提升我国海底地名提案水平，国家海洋信息中心对《海底地名命名标准》（B-6出版物）历年版本均进行了翻译，2013年经国际海道测量局（IHB）授权，《海底地名命名标准》（B-6出版物4.1.0）英文/中文版在中国出版发行，并在国际海道测量组织（IHO）的官方网站上发布。

目前，国家海洋信息中心组织相关人员在对《海底地名命名标准》（B-6出版物4.1.0）英文/中文版编译成果认真研究的基础上，完成了对该标准4.2.0版的编译。本书翻译过程中得到了自然资源部国际合作司的大力支持，民政部地名研究所、自然资源部第二海洋研究所、广州海洋地质调查局、中国大洋矿产资源研究开发协会等单位专家提出了宝贵意见。SCUFN中国委员，国家海洋信息中心副总工程师李四海研究员对全书进行质量把控，并与SCUFN积极沟通，获得出版许可。中国大洋矿产资源研究开发协会原秘书长、中国常驻国际海底管理局原副代表毛彬研究员对全书做了审校。在此一并表示感谢！

由于译者水平有限，不妥之处在所难免，欢迎指正。

译者

2023年11月

INTERNATIONAL HYDROGRAPHIC
ORGANIZATION

国际海道测量组织

INTERGOVERNMENTAL
OCEANOGRAPHIC
COMMISSION

政府间海洋学委员会

STANDARDIZATION
OF UNDERSEA
FEATURE NAMES

海底地名命名标准

GUIDELINES

指导原则

PROPOSAL FORM

海底地名提案表

TERMINOLOGY

术语

Publication B-6

B-6出版物

Edition 4.2.0–October 2019

4.2.0版，2019年10月

English/Chinese Version

英文/中文版

Published by the
INTERNATIONAL HYDROGRAPHIC
ORGANIZATION

出版者
国际海道测量组织

Foreword

The Guidelines, the Name Proposal Form and the List of Terms and Definitions contained in the IHO-IOC publication B-6 "Standardization of Undersea Feature Names" were originally developed through collaboration between the "GEBCO Sub-Committee on Undersea Feature Names" appointed by the "Joint IHO-IOC Guiding Committee for GEBCO (GGC)", and the Working Group on Undersea and Maritime Features of the "United Nations Group of Experts on Geographical Names (UNGEGN)" in accordance with provisions of appropriate resolutions of United Nations Conferences on the Standardization of Geographical Names (UNCSGN). The UNGEGN Working Group on Undersea and Maritime Features was disbanded in 1984 but a liaison has been maintained between IHO and UNGEGN to facilitate communication and cooperation.

This edition 4.2.0 of the English/Chinese version of B-6 supersedes the previous edition published by the IHO in 2013 (updated February 2017). Other versions of this edition are, or will be also available in English/Spanish, English/Russian, English/Japanese, English/Korean and English/Portuguese.

At the request of the "Joint IHO-IOC Guiding Committee for GEBCO", in order to obtain the largest distribution of these Guidelines and to bring the Geographical Names of Undersea Features to a better Standardization, the B-6 is available gratis in digital form from the IHO website (www.iho.int) and GEBCO website (www.gebco.net).

前 言

国际海道测量组织—政府间海洋学委员会（IHO-OC）B-6出版物"海底地名命名标准"刊登的海底地名命名指南、海底地名提案表、术语和定义，最初由国际海道测量组织—政府间海洋学委员会（IHO-IOC）全球通用大洋水深制图计划（GEBCO）联合指导委员会（GGC）任命的"GEBCO海底地名分委会（SCUFN）"和"联合国地名专家组（UNGEGN）"之海底和海洋地理实体工作组依据联合国地名标准化会议（UNCSGN）相关决议条款共同协商制定。UNGEGN海底和海洋地理实体工作组于1984年解散，但为了便于沟通与合作，国际海道测量组织（IHO）和联合国地名专家组（UNGEGN）一直保持着联系。

B-6出版物4.2.0英文/中文版替代了先前由国际海道测量组织（IHO）出版的2013年版（2017年2月更新）。另外，本出版物还有或即将有英文/西文、英文/俄文、英文/日文，英文/韩文和英文/葡文版本。

根据国际海道测量组织—政府间海洋学委员会（IHO-IOC）全球通用大洋水深制图计划（GEBCO）指导委员会的要求，为了使这些指导原则得到广泛的传播与应用，促进海底地理实体的命名更加标准化，用户可从国际海道测量组织官方网站（www.iho.int）和全球通用大洋水深制图计划（GEBCO）官方网站（www.gebco.net）免费获取B-6出版物电子版文件。

CONTENTS 目录

- Standardization of Undersea Feature Names—Introduction
  海底地名命名标准—引言 ··· 1
- Guidelines for the Standardization of Undersea Feature Names
  海底地名命名标准指南 ·· 4
  - Ⅰ. General
    Ⅰ. 总则 ··· 4
  - Ⅱ. Principles for Naming Features
    Ⅱ. 命名原则 ·· 5
  - Ⅲ. Procedures for Naming Features
    Ⅲ. 命名程序 ·· 7
- Undersea Feature Name Proposal—English/Chinese version
  海底地名提案表—英文/中文版 ·· 9
- Terminology—Notes
  术语—说明 ·· 12
- Undersea Feature Terms and Definitions
  海底地理实体通名及其定义 ·· 13
- Chinese Alphabetical Index of the Chinese Terms Shown in the foregoing List of "UNDERSEA FEATURE TERMS AND DEFINITIONS", with Cross-References to the English Terms
  "海底地理实体通名及其定义"中按汉语拼音字母排序的通名索引表—英文/中文版
  ··· 23
- Appendix A: User's Guide for Preparation of Undersea Feature Name Proposals to the GEBCO Sub-Committee on Undersea Feature Names (SCUFN)
  附录 A：全球通用大洋水深制图计划（GEBCO）海底地名分委会（SCUFN）海底地名提案编制用户指南 ··· 26
- Appendix B: Procedure for the Adoption of Undersea Feature Names Proposed by National Geographical Naming Authorities (*Fast-Track Procedure*)
  附录 B：国家地名管理机构提交的海底地名提案的通过程序（快速审议程序）··· 34

List of Acronyms

GEBCO: General Bathymetric Chart of the Oceans
IBC: International Bathymetric Chart
IHO: International Hydrographic Organization
IOC: Intergovernmental Oceanographic Commission (UNESCO)
SCUFN: GEBCO Sub-Committee on Undersea Feature Names
UN: United Nations
UNCLOS: United Nations Convention on the Law of the Sea
UNESCO: United Nations Educational, Scientific and Cultural Organization
UNCSGN: United Nations Conference on the Standardization of Geographical Names
UNGEGN: United Nations Group of Experts on Geographical Names

缩略语一览表

GEBCO： 全球通用大洋水深制图计划
IBC： 国际海底测深图
IHO： 国际海道测量组织
IOC： 政府间海洋学委员会（UNESCO）
SCUFN： 海底地名分委会（全球通用大洋水深制图计划）
UN： 联合国
UNCLOS： 联合国海洋法公约

UNESCO： 联合国教科文组织

UNCSGN： 联合国地名标准化会议

UNGEGN： 联合国地名专家组

Standardization of Undersea Feature Names

INTRODUCTION

1. In past years, considerable concern has been expressed at the indiscriminate and unregulated naming of undersea features which often get into print in articles submitted to scientific publications, or on maps and charts, without any close scrutiny being made concerning their suitability, or even whether the feature has already been discovered and named.

In order to remedy this situation and to bring the geographical names of undersea features to a better standardization, the IHO, at its XIII th I.H. Conference (May 1987) and the IOC, at its 14th Assembly (March 1987) adopted similar motions on this subject, the substance of which is recalled below.

i) Marine scientists and others wishing to name undersea features, are strongly encouraged to check their proposals with the Gazetteer of Undersea Feature Names on the GEBCO website (www.gebco.net) taking into account the guidelines contained in this publication (B-6), including the use of the "Undersea Feature Name Proposal" form contained herein, and to submit all proposed new names for clearance, either to their appropriate national authority, or, where no such national authority exists, to the IHO (info@iho.int) or IOC (info@unesco.org), for consideration by the "GEBCO Sub-Committee on Undersea Feature Names" (SCUFN), which may advise on any potentially confusing duplication of names.

ii) Publishers of maps, and editors of scientific publications, in their country, are invited to require compilers and authors to provide written evidence of such clearance before accepting for publication any maps or articles containing new names for undersea features.

2. In 2008, new Terms of Reference for the "GEBCO Sub-Committee on Undersea Feature Names" (SCUFN) were adopted by IHO and IOC, including the following:

海底地名命名标准

引言

1. 在过去的几年中，混乱无章的海底地名经常出现在某些公开出版的学术刊物的论文中或地图和海图上，而对这些地名的适用性，甚至对该实体是否已被发现和命名都未做仔细的审查。这种乱象已经引起了人们的广泛关注。

为了纠正上述情况，使海底地名命名更趋标准化，国际海道测量组织（IHO）于其第13届大会（1987年5月）上，政府间海洋学委员会（IOC）于其第14届大会（1987年3月）上分别就这一问题通过了相似的决议，其主要内容如下。

ⅰ）强烈鼓励有意进行海底地名命名的海洋科学家和其他人员，首先将拟命名海底地名提案与全球通用大洋水深制图计划（GEBCO）官方网站（www.gebco.net）发布的海底地名词典相对照，并参考本文件（B-6出版物）中阐述的指导原则，包括使用《海底地名提案表》，再将提案提交至本国有关机构审核以获得许可。如果本国没有相关机构，则可将提案提交至国际海道测量组织（IHO）（info@iho.int）或政府间海洋学委员会（IOC）（info@unesco.org），由GEBCO海底地名分委会（SCUFN）审议研究后，对提案是否存在混淆和重复等问题提出意见。

ⅱ）各国地图出版者和科学出版物编辑，在同意出版含有新海底地名的图件或文章之前，应要求编者或作者提供所用新海底地名已获得上述有关机构许可的书面证明。

2. 2008年国际海道测量组织（IHO）和政府间海洋学委员会（IOC）通过了GEBCO海底地名分委会（SCUFN）新的职责，包括内容如下：

i) It is the function of the Sub-Committee to select those names of undersea features in the world ocean appropriate for use on GEBCO graphical and digital products, on the IHO small-scale INTernational chart series, and on the regional International Bathymetric Chart (IBC) series.

ii) The Sub-Committee shall:

- select undersea feature names from:

    – names provided by national authorities and international organizations concerned with nomenclature;

    – names submitted to the Sub-Committee by individuals (with the exception of SCUFN members), agencies and organizations involved in marine research, hydrography, etc.;

    – names appearing in scientific publications or on appropriate charts and maps.

    – names submitted to the Sub-Committee by the Chairpersons or Chief Editors of International Bathymetric Chart projects, in relation to the work on these projects.

    All selected names shall adhere to the principles contained in this publication and be supported by valid evidence. Such names shall be reviewed before they are added to the Gazetteer.

- define when appropriate the extent of named features;

- provide advice to individuals and appropriate authorities on the selection of undersea feature names located outside the external limits of the territorial sea and, on request, inside the external limit of the territorial sea;

- encourage the establishment of national authorities concerned with the naming of undersea features when such authorities do not exist;

ⅰ）分委会的职责是选定适用于GEBCO的图件和数字产品、IHO小比例尺国际系列海图（INT）和区域性系列国际海底测深图（IBC）的世界大洋海底地名。

ⅱ）分委会应：

- 从下列渠道选定海底地名：

    – 与命名有关的国家机构和国际组织提供的地名；

    – 与海洋研究和海道测量等相关的个人（SCUFN成员除外）、机构和组织提交给分委会的地名；

    – 在科学出版物或者适当的海图和地图上出现的地名；

    – 国际水深制图项目负责人或主编，因开展项目相关工作提交给分委会的地名。

    所有选定的地名都应与本出版物中规定的原则相一致，并提供有效的支持证据。这些地名经审查后才能加入海底地名词典。

- 对命名的地理实体范围进行正确的界定；

- 针对提交位于领海外部界限以外的海底地名的选择，或依据要求提交领海外部界限以内的海底地名的选择，向个人和有关机构提供咨询意见；

- 鼓励尚未设立海底地名管理机构的国家设立相关机构；

- prepare and maintain the GEBCO Gazetteer;

- encourage the use of undersea feature names included in the GEBCO Gazetteer, on any maps, charts, scientific publications, and documents by promulgating these names widely;

- prepare and maintain this publication and encourage its use;

- review and address the need for revised or additional terms and definitions for undersea features;

- maintain close liaison with the UNGEGN, the focal point of which shall be invitations to attend meetings of the Sub-Committee, and with international or national authorities concerned with the naming of undersea features;

- provide, where feasible, historical information regarding the origin of pre-existing published names and historical variant names. This research will include discovery ship and/or organization, information regarding the individual or vessel being commemorated or geographic feature with which the name is associated, origin of variant names if required and source material regarding naming information.

- 编制和维护GEBCO海底地名词典；

- 鼓励在各种地图、海图、科学出版物以及文献中使用GEBCO海底地名词典中的地名，使它们得以广泛传播；

- 编制和维护本出版物并推广使用；

- 审议和处理海底地理实体通名修订和新增通名和定义的问题；

- 与联合国地名专家组（UNGEGN）、被邀请参加海底地名分委会（SCUFN）会议的联系人，以及与海底地名命名相关的国际组织或国家机构保持密切联系；

- 在可能的情况下，提供已公布的海底地名来源和更名的历史信息，包括发现该地理实体的调查船舶和/或执行该调查的机构，为纪念个人或调查船而命名的信息，或因为相关地理特征而命名的信息等。如必要还需提供更名原因以及有关命名的原始资料。

Guidelines for the Standardization of Undersea Feature Names

海底地名命名标准指南

I. General

I. 总则

A. International concern for naming undersea features is limited to those features entirely or mainly (more than 50%) outside the external limits of the territorial sea, not exceeding 12 nautical miles from the baselines, in agreement with the *United Nations Convention on the Law of the Sea*.

A. 国际上关注的海底地理实体命名主要限于其全部或主体（＞50%）位于领海以外的海底地理实体。根据《联合国海洋法公约》规定，领海宽度应从本国领海基线起算不超过12海里。

B. "Undersea feature" is a part of the ocean floor or seabed that has measurable relief or is delimited by relief.

B. "海底地理实体"是洋床或海底的一部分，其具有可测量的地形起伏或由地形起伏划定其边界范围。

C. Names used for many years may be accepted even though they do not conform to normal principles of nomenclature. Existing names may be altered to avoid confusion, remove ambiguity or to correct spelling.

C. 已使用多年的地名即使不符合现行命名标准也应接受，但为了避免混乱，这些地名需要做适当修改，或去掉模糊不清的成分，或更改其拼写方式。

D. Names approved by national authorities in waters beyond the territorial sea should be accepted by other States if the names have been applied in conformance with internationally accepted principles. Names applied within the territorial sea of a State should be recognized by other States.

D. 由一国海底地名管理机构批准的位于领海以外的海底地名，如符合现有的国际标准，其他国家应该接受。一个国家在其领海以内命名的海底地名，其他国家应予以认可。

E. In the event of a conflict, the persons and/or authorities involved should resolve the matter. Where two names have been applied to the same feature, the older name generally should be accepted. Where a single name has been applied to two different features, the feature named first generally should retain the name.

E. 当出现命名冲突时，相关机构和/或个人应协商解决。当一个地理实体有两个不同名称时，一般应保留使用较早的那个名称；当同一个名称用于两个不同的地理实体时，先命名的地名应予以保留。

F. Names not in the writing system of the country applying the names on maps or other documents should be transliterated according to the system adopted by the appropriate national authority applying the names.

F. 在地图或其他文献中使用非本国拼写方式的海底地名时，应以本国地名管理机构认可的拼写方式进行译写。

G. In international programmes, it should be the policy to use forms of names applied by national authorities having responsibility for the pertinent area.

G. 作为一种政策来规定，在国际计划中，有关区域的地名应使用负责该区域的国家机构使用的名称。

H. States may utilize their preferred versions of exonyms.

II. Principles for Naming Features

Note: a specific term followed by a generic term make up a feature name.

A. Specific terms

1. Short and simple specific terms are preferable.

2. The principal concern in naming is to provide effective, conveniently usable, and appropriate reference; commemoration of persons or ships is a secondary consideration.

3. The first choice of a specific term, where feasible, should be one associated with a geographical feature; e.g. Aleutian Ridge, Mariana Trench, Katsuura Canyon.

4. Other choices for specific terms can commemorate ships or other vehicles, expeditions or scientific institutes involved in the discovering and/or delineation of the feature, or to honour the memory of famous persons, preferably personalities whose contribution to ocean sciences, exploration or history has been internationally recognized. Where a ship name is used, it should be that of the discovering ship, or if that has been previously used for a similar feature, it should be the name of the ship verifying the feature, e.g. San Pablo Seamount, Atlantis II Seamounts.

5. Names of living persons will normally not be accepted, in accordance with the recommendation in the UNCSGN Resolution VIII/2. In the rare cases where names of living persons are used (surnames are preferable), they will be limited to those who have made an outstanding or fundamental contribution to ocean sciences.

H. 对于外来地名，使用国可以自行选择使用版本。

II. 命名原则

说明： 海底地名由专名和通名组成，专名在前，通名在后。

A. 专名

1. 专名应尽量简短。

2. 专名采词的首要原则是实用性，即方便使用且具有适当的参照作用，其次考虑纪念名人或船只。

3. 在可能情况下，专名采词时应首先考虑邻近的地理实体名称，如阿留申海脊、马里亚纳海沟、胜浦海底峡谷。

4. 专名采词也可用来纪念发现和/或确定地理实体的船只或其他运载工具、考察探险或科学机构，或者纪念在海洋科学研究、科学考察或历史文化研究等方面得到国际认可的杰出人物。采用船名命名时，该船应是发现该地理实体的船只，如果该船名先前已用于命名其他地理实体，那么该船必须是调查并核实该地理实体的船只。用调查船名命名的海底地名，如圣·巴勃罗海山、阿特兰蒂斯II海山。

5. 根据联合国地名标准化会议（UNCSGN）VIII/2决议的建议，一般不以在世人名来命名海底地理实体。但在极少数情况下，需要以在世人名命名（最好用姓氏）时，此人必须为海洋科学做出过杰出或重要的贡献。

6. Groups of like features may be named collectively for specific categories of historical persons, mythical features, stars, constellations, fish, birds, animals, etc. For example:

6. 特征相似的群组地理实体的命名可以采用某类名称的集合来进行群组化命名，如历史人物、神话传说、星体、星座、鱼类、鸟类、动物等。例如：

Musicians Seamounts
音乐家海山群
{
Mozart Seamount
莫扎特海山
Brahms Seamount
布拉姆斯海山
Schubert Seamount
舒伯特海山
}

Grouping of like features in categories should be determined for distinct geographical configurations, based on considerations of the (their) morphological, tectonic, or structural domain. Some examples are a series of features forming a single line (e.g. Emperor Seamount Chain), or a concentration of features in a certain geographical domain (e.g. Great Writers Seamount Province, Parece Vela Fracture Zone Province). In the case of names in the vicinity of Antarctica, it is recommended that specific terms should relate to the Antarctic region, explorers, researchers or vessels.

对于成组的相似地理实体，应基于对地理实体（群组）的形态、构造、结构范围的分析，根据其不同的地理属性确定群组地理实体的分组。例如，由一系列呈线状排列的地理实体形成的群组（如皇帝海山链），或若干同类型地理实体聚集在一定范围内，形成一个特征区（如文豪海山特征区、帕里西维拉断裂带特征区）。对南极洲附近的海底地理实体命名时，建议选取与南极区域、探险者、研究人员或船只等有关的词语。

7. Descriptive names are acceptable, particularly when they refer to distinguishing characteristics (i.e. Hook Ridge, Horseshoe Seamount). However, this is only advised when a characteristic shape has been established by definitive topographic exploration.

7. 可采用形象化词语命名，尤其是形象特征明显的实体（即鱼钩海脊、马蹄海山）。但建议此方法仅适用于通过地形测量对地理实体形态确认无疑的情况。

8. Names of well-known or large features that are applied to other features should have the same spelling.

8. 将知名的或者大型地理实体名称应用到对其他地理实体命名时，名称拼写应该保持一致。

9. A specific term should not be translated from the language of the nation providing the accepted name.

9. 已经被收录的专名可保留原命名国的语言形式，不必翻译成其他语言。

B. Generic terms

1. Generic terms should be selected from the following list of definitions to reflect physiographic descriptions of features. This list, along with images illustrating the generic terms, can also be found on the following website: www.scufnterm.org.

2. Generic terms applied to features appearing on charts or other products should be in the language of the nation issuing the products. In cases where terms have achieved international usage in a national form, that form should be retained.

3. It should be recognized that as ocean mapping continues, features will be discovered for which existing terminology is not adequate. New terms required to describe those features should conform to this publication.

III. Procedures for Naming Features

A. Individuals and agencies applying names to unnamed features located outside the external limit of the territorial sea should adhere to internationally accepted principles and procedures, as detailed in this publication.

B. New proposals should be submitted on an "Undersea Feature Name Proposal" form as contained in this publication. "User's Guide for Preparation of Undersea Feature Name Proposals" is provided at Appendix A.

C. Prior to the naming of a feature, its character, extent, and position should be established sufficiently for identification. Positions (point, line or polygon) should be given as geographic coordinates, preferably in shape format.

B. 通名

1. 通名应从下列反映海底地理实体自然地理特征的通名定义列表中选取。该列表及其图文描述也可从该网站上查询：www.scufnterm.org。

2. 在海图和其他产品上标注海底地理实体的通名时，应该采用出版国语言。如果使用某国语言形式的通名已经得到国际上的广泛使用，应予以保留。

3. 应该认识到，随着海洋制图工作的不断发展，可能发现现有海底地理实体定义的术语已经不合适，需要引入新的通名对地理实体进行重新定义，但这些通名应符合本出版物的要求。

III. 命名程序

A. 个人或机构为位于领海外部界限以外的尚未命名的海底地理实体命名时，应遵循本出版物中规定的国际公认的原则和程序。

B. 新的地名提案应以本出版物中所包含的"海底地名提案表"形式提交。附录A中提供了"海底地名分委会（SCUFN）海底地名提案编制指南"。

C. 对海底地理实体命名之前，应先准确测定该实体的特征、范围和位置，以便于识别。位置（点、线或多边形）采用地理坐标表示，最好采用矢量文件格式。

D. There is significant benefit to be gained from mutual consultation by all interested parties in preparing and submitting proposals to SCUFN. National naming authorities are encouraged to consult on undersea features names in their mutual areas of interest prior to submitting proposals to SCUFN.

E. Where no appropriate national authority exists, clearance should be sought through either the IHO Secretariat or the IOC Secretariat, as indicated on the "Proposal Form".

F. If a national authority decides to change either the specific or generic term of a feature it named originally, information explaining the reason for the change should be circulated to other authorities. If there is opposition to a name change, the involved authorities should communicate with each other to agree on a solution.

G. National authorities approving names of features should regularly publicize their decisions. Under certain conditions, for example for names that are in long term use and appear on published charts, a national naming authority may submit a set of names for adoption en bloc by SCUFN through a fast-track procedure, as described at Appendix B.

H. National authorities naming features within their territorial sea should conform to the principles and procedures stated above.

D. 在编制和向海底地名分委会（SCUFN）提交海底地名提案时，所有利益相关方相互协商将颇有裨益。我们鼓励各国地名管理机构在提交位于他国共同感兴趣区域的提案前，与有关国家相互协商后再向海底地名分委会（SCUFN）提交。

E. 如果一个国家没有设置相应的地名管理机构，应按"海底地名提案表"所示，通过国际海道测量组织（IHO）或者政府间海洋学委员会（IOC）秘书处获得海底地名提案的许可。

F. 当一国地名管理机构决定修改已命名的地理实体的专名或通名时，应该将更名原因等信息通报给其他有关国家地名管理机构。如产生分歧时，相关各方应相互沟通以商定解决办法。

G. 批准海底地名的国家地名管理机构应该定期向公众公布其决定。在一定条件下，例如海底地名已被长期使用并标注于公开出版的海图中，该国地名管理机构可通过快速审议程序批量提交此类地名，供海底地名分委会（SCUFN）进行整体审议通过，具体见附录B。

H. 国家地名管理机构在对本国领海以内的海底地理实体命名时，也应遵循上述原则和程序。

INTERNATIONAL HYDROGRAPHIC ORGANIZATION	INTERGOVERNMENTAL OCEANOGRAPHIC COMMISSION (of UNESCO)
国际海道测量组织（IHO）	政府间海洋学委员会（IOC）

Undersea Feature Name Proposal——English/Chinese version
海底地名提案表——英文/中文版
(See **NOTE** 详见说明)

Note:
说明：
a) Translation in Chinese is provided for convenience. However, the form should be filled in English.
本表提供的中文译文仅供提案编制参考，对外提交提案表须用英语填写。
b) The boxes will expand as you fill the form.
可根据内容调整表格。
c) Please apply guidelines in Appendix A.
填表时应遵循附录A中的条款。

Name Proposed: 拟命名：			

Geometry that best defines the feature (Yes/No)： 最佳界定拟命名海底地理实体的几何图形（是/否）：						
Point 点	Line 线	Polygon 多边形	Multiple points 多点	Multiple lines* 多线*	Multiple polygons* 多个多边形*	Combination of geometries* 多种几何图形组合*

* Geometry should be clearly distinguished when providing the coordinates below.
*根据下列坐标，可清楚地识别地理实体的几何图形。

	Lat. (e.g. 63°32.6′ N) 纬度（如63°32.6′ N）	Long. (e.g. 046°21.3′ W) 经度（如046°21.3′ W）
Coordinates**: 坐标**：		

** For quality control and to minimize risks of making errors, it is recommended to provide proposals in digital format (pdf) as well as geometry (point, line, ...) files in shape format.
** 为控制质量和尽量减少错误，建议以电子版形式提交提案表（PDF）和以矢量格式提供几何（线、点）文件。

Feature Description: 拟命名海底地理实体描述：	Maximum Depth: 最大水深：			
	Minimum Depth: 最小水深：		Shape: 形状：	
	Total Relief: 总起伏：		Dimension/Size: 尺度/大小范围：	

Associated Features: 相关地理实体：	

Chart/Map References: 参照的海图/地图：	Shown Named on Map/Chart: 标有该地理实体及名称的地图/海图：	
	Shown Unnamed on Map/Chart: 只标有该地理实体但未标出其名称的地图/海图：	
	Within Area of Map/Chart: 包含拟命名地理实体区域的地图/海图：	

Reason for Choice of Name (if a person, state how associated with the feature to be named): 选择名称的理由（如果是人名，应说明与拟命名地理实体的关系）：	

Discovery Facts: 发现事实：	Discovery Date: 发现日期：	
	Discoverer (Individual, Ship): 发现者（个人、船只）：	

Supporting Survey Data, including Track Controls: 支持获得本次发现的调查资料，包括测线控制：	Date of Survey: 调查日期：	
	Survey Ship/Platform: 调查船/平台：	
	Sounding Equipment: 测深设备：	
	Positioning System: 定位系统：	
	Estimated Horizontal Accuracy, in nautical miles (M): 估计水平精度（海里）：	
	Survey Track Spacing: 测线间隔：	
	Supporting material can be submitted as Annex in analog or digital form. 支持材料可作为附件以图件或数字形式提交。	

Proposer(s): 提案人：	Name(s): 姓名：	
	Date: 日期：	
	E-mail: 电子信箱：	
	Organization and Address: 单位和地址：	
	Concurrer (name, e-mail, organization and address): 共同提案人（姓名、电子邮箱、单位和地址）：	

Remarks: 备注：	

NOTE: This form should be forwarded, when completed:
说明：此表应完整填写后提交：

a) If the undersea feature is located inside the external limit of the territorial sea:
– to your "National Authority for Approval of Undersea Feature Names" or, if this does not exist or is not known, either to the IHO or to the IOC (see addresses below):

如果拟命名的海底地理实体位于领海的外部界限以内：

– 向本国的国家海底地名管理机构提交；如尚未设立国家海底地名管理机构或不清楚有无此类机构，可向国际海道测量组织（IHO）或政府间海洋学委员会（IOC）提交。见下列地址；

b) If at least 50% of the undersea feature is located outside the external limits of the territorial sea:
– to the IHO or to the IOC, at the following addresses:

如果拟命名的海底地理实体至少50%的部分位于领海外部界限以外：

– 向国际海道测量组织（IHO）或政府间海洋学委员会（IOC）提交。见下列地址：

Internationale Hydrographic Organisation (IHO) 4, Quai Antoine 1er B.P. 445 MC 98011 MONACO CEDEX Principality of MONACO Fax: +377 93 10 81 40 E-mail: info@ihb.mc Web: www.iho.int	Intergovernmental Oceanographic Commission (IOC) UNESCO Place de Fontenoy 75700 PARIS France Fax: +33 1 45 68 58 12 E-mail: info@unesco.org Web: http://ioc-unesco.org/

Terminology

NOTES (See also "FOREWORD")

The list in Section I hereafter "GENERIC TERMS" is comprised of terms that are defined as closely as possible to correspond to their usage in references appearing in the literature of ocean science, hydrography and exploration. In developing the definitions, it was realized that modern investigations at sea have the advantage of using very advanced instrumentation and technology that enables a more precise description of certain features than was previously possible. This has sometimes lead to finding that historically named features, do not physically exist. There has also been an attempt to limit the usage of precise physical dimensions in the definition of features. In preference, words that indicate relative sizes such as extensive, large, limited and small have been used. The definitions are based almost exclusively on a geomorphological description of the features themselves, although some terms with implications on the origin or composition of features are also included. The terms in this list must not be construed as having any legal or political connotation whatsoever. Nor do they necessarily conform to the hydrographic/navigation usage as appearing in the Hydrographic Dictionary (IHO Publication S-32).

The list in Section II hereafter "GENERIC TERMS USED FOR HARMONIZATION WITH OTHER GAZETTEERS", is comprised of terms no longer used in modern physiographic terminology but which appear for some features in the GEBCO Gazetteer and/or in other gazetteers. They are kept in this publication to facilitate harmonization between gazetteers, and also to recognize that generic terms in some named features, such as "cap" or "pass", have widely accepted longtime usage. However, they are considered obsolete and their use is not recommended for new feature names.

For terms in the list having no definition, an alternative and recognized generic term is provided.

术语

说明（见"前言"）

下文第一节中列出的"通名"在定义上尽量与出现在海洋科学、海道测量学和海洋调查文献中的用法保持一致。在制定这些定义时，我们已经认识到现代海洋调查具有运用先进仪器和技术的优势，从而能够比以前更加精确地测量海底地理实体。这项工作使我们发现有些历史上已命名的海底地理实体实际上并不存在。一直以来人们通常不用精确的地理空间范围定义海底地理实体，而偏向使用相对大小或模糊性的词语，如广阔的、大的、有限的、小的等。虽然包括一些反映地理实体成因或成分含义的通名，但这些定义几乎完全是基于对地理实体本身地貌形态的描述。本清单中的通名不具有任何法律或政治含义，也无须与海道测量词典（IHO S-32出版物）中出现的海道测量/航行用词完全一致。

下文第二节"与其他词典兼容的通名"包括在现代自然地理学中已不再使用的通名，但仍出现在全球通用大洋水深制图计划（GEBCO）地名中和/或其他地名词典中。本出版物保留这些通名，目的是促进不同地名词典之间的相互兼容。需要指出的是，一些已长期使用且被广泛接受的通名，如"海角"和"隘口"，由于其在现代自然地理学术语中已不再使用，因此不建议用于对新发现的海底地理实体命名中。

对于清单中没有给出定义的通名，本出版物提供了另一个被广泛认可的通名作为替代。

Undersea Feature Terms and Definitions

Notes:

1) Terms written in capitals in the definitions are themselves defined elsewhere in the list at sections Ⅰ and Ⅱ.

2) The plural form of a generic term may be used to represent a closely associated group of features of the same generic type (e.g. SEAMOUNTS).

3) Generic terms for features that have a genetic implication are marked with an asterisk (*). Name proposals that contain a generic term with genetic implications must include geological and/or geophysical evidence as well as bathymetric data.

4) Examples of images illustrating the generic terms listed below can be found on the following website: www.scufnterm.org.

Ⅰ. GENERIC TERMS

NOTE:

Only the generic terms in this section should be used in any new undersea feature name proposal that is intended for submission to SCUFN.

ABYSSAL PLAIN

An extensive, flat or gently sloping region, usually found at depths greater than 4,000 m.

APRON

A gently dipping SLOPE, with a smooth surface, commonly found around groups of islands and SEAMOUNTS.

BANK

An elevation of the seafloor, at depths generally less than 200 m, but sufficient for safe surface navigation, commonly found on the continental shelf or near an island.

海底地理实体通名及其定义

说明：

1）定义中的粗体字通名在本部分第一和第二节其他地方予以定义。

2）通名的复数形式常用于表示一组紧密关联且同类地理实体群（如海山群）。

3）对具有成因含义的地理实体通名标注了星号（*）。含有这类通名的地名提案在提供水深资料的同时，须提交地质和/或地球物理方面的证据。

4）下文列出的通名及其图文描述可在该网站中查询：www.scufnterm.org。

Ⅰ.通名

说明：

任何拟提交至海底地名分委会（SCUFN）的海底地名提案只可使用本节中列出的通名。

深海平原

范围广阔、地势平坦或坡度平缓的海底区域，一般见于水深大于4 000 m的区域。

冲积裙

地形平缓倾斜、表面平滑的海底**斜坡**，一般见于群岛和**海山群**周围。

滩

海底高地，上覆水深一般浅于200 m，但可满足海面安全航行，常见于大陆架上或海岛附近。

BASIN

A depression more or less equidimensional in plan and of variable extent.

CALDERA*

A roughly circular, cauldron-like depression generally characterized by steep sides and formed by collapse, or partial collapse, during or following a volcanic eruption.

CANYON

An elongated, narrow, steep-sided depression that generally deepens down-slope.

DEEP

A localized depression within the confines of a larger feature, such as a TROUGH, BASIN or TRENCH.

ESCARPMENT

An elongated, characteristically linear, steep slope separating horizontal or gently sloping areas of the seafloor.

FAN

A relatively smooth, depositional feature continuously deepening away from a sediment source commonly located at the lower termination of a CANYON or canyon system.

FRACTURE ZONE*

A long narrow zone of irregular topography formed by the movement of tectonic plates associated with an offset of a spreading ridge axis, characterized by steep-sided and/or asymmetrical RIDGES, TROUGHS or ESCARPMENTS.

GAP

A narrow break in a RIDGE, RISE or other elevation. Also called PASSAGE.

GUYOT

A SEAMOUNT with a comparatively smooth flat top.

海盆

海底洼地，平面大体呈等维分布，范围大小不一。

塌陷火山口*

海底火山喷发时或喷发后，发生塌陷或部分塌陷形成的洼地，整体呈圆形，通常具有陡峭的边坡。

海底峡谷

形态狭长且边坡陡峭的海底洼地，一般顺陆坡而下，谷底不断变深。

海渊

在**海槽**、**海盆**或**海沟**等大型地理实体内部出现的局部洼地。

海底崖

呈线状延伸展布的海底陡坡，将地形平坦或平缓倾斜的区域切断。

海扇

地形相对平缓的海底沉积体，通常位于**海底峡谷**或海底峡谷系的底部，水深随着远离沉积源而不断加大。

断裂带*

伴有海底扩张脊轴错断的板块构造运动而形成的形状不规则的狭长地带，内部常见有两翼陡峭和/或不对称的**海脊**、**海槽**或**海底崖**等。

裂谷

海脊、**海隆**或其他海底高地中出现的狭窄断裂，也叫山口。

平顶海山

顶部地形相对平坦的**海山**。

HILL

A distinct elevation generally of irregular shape, less than 1,000 m above the surrounding relief as measured from the deepest isobath that surrounds most of the feature.

HOLE

A depression of limited extent with all sides rising steeply from a relatively flat bottom.

KNOLL

A distinct elevation with a rounded profile less than 1,000 m above the surrounding relief as measured from the deepest isobath that surrounds most of the feature.

LEVEE

A depositional embankment bordering a CANYON, VALLEY or SEA CHANNEL.

MOAT

An annular or partially annular depression commonly located at the base of SEAMOUNTS, islands and other isolated elevations.

MOUND*

A distinct elevation with a rounded profile generally less than 500 m above the surrounding relief as measured from the deepest isobath that surrounds most of the feature, commonly formed by the expulsion of fluids or by coral reef development, sedimentation and (bio) erosion.

MUD VOLCANO*

A MOUND or cone-shaped elevation formed by the expulsion of non-magmatic liquids and gasses.

PEAK

A conical or pointed elevation on a larger feature such as a SEAMOUNT.

海丘

轮廓清晰可辨的海底隆起区，一般形状不规则，从环绕其主体最大等深线起算，至顶部最大高差小于1 000 m。

海穴

海底出现的局部洼地，四周边坡从地形较平坦的底部急剧上升。

圆丘

轮廓清晰可辨的海底隆起区，呈圆形，从环绕其主体的最大等深线起算，至顶部最大高差小于1 000 m。

海堤

自然沉积而成的坝体，一般位于**海底峡谷**、**海谷**或**海底水道**的边缘。

环形洼地

环形或部分呈环形的海底洼地，通常位于**海山群**、群岛和其他孤立海底高地群的基底处。

矮丘*

轮廓清晰可辨的海底隆起区，呈圆形。从环绕其主体的最大等深线起算，至顶部最大高差一般小于500m。通常由海底流体喷发形成，或由珊瑚礁发育、沉积和（生物）侵蚀作用形成。

泥火山*

由非岩浆流体或气体喷发而形成的**矮丘**或圆锥状隆起。

海底峰

位于**海山**等较大型地理实体上的圆锥状或尖顶状高地。

PINNACLE

A spire-shaped pillar either isolated or on a larger feature.

PLATEAU

A large, relatively flat elevation that is higher than the surrounding relief with one or more relatively steep sides.

PROVINCE

A geographically distinct region with a number of shared physiographic characteristics that contrast with those in the surrounding areas. This term should be modified with the generic term that best describes the majority of features in the region, e.g. "SEAMOUNT" in "Baja California SEAMOUNT PROVINCE".

REEF*

A shallow elevation composed of consolidated material that may constitute a hazard to surface navigation.

RIDGE

An elongated elevation of varying complexity and size, generally having steep sides.

RIFT*

An elongated depression bounded by two or more faults formed as a breach or split between two bodies that were once joined.

RISE

A broad elevation that generally rises gently and smoothly from the surrounding relief.

SADDLE

A broad pass or col in a RIDGE, RISE or other elevation.

尖礁

尖顶状岩柱，或孤立出现，或位于较大型地理实体上。

海底高原

大范围且地形相对平坦的海底高地，地势明显高出周围区域，一侧或多侧边坡地形陡峭。

特征区

具有共同地理特征且与周边形成清晰对比的地理区域。本通名可依据最能反映该区域主要地理实体特征的通名进行相应调整，如"下加利福尼亚**海山**特征区"中的"**海山**"。

礁*

由固结物质组成的浅水隆起区，可能会危及海面航行安全。

海脊

大小和范围变化复杂的狭长隆起地带，通常具有陡峭的边坡。

断裂谷*

由两个及以上断层包围而成的狭长洼地，是两个曾经连在一起的地块分离形成的。

海隆

相对于周围地形平缓抬升且宽阔的海底隆起区。

鞍部

海脊、**海隆**或其他隆起区中发育的宽阔通道或隘口。

SALT DOME*

A distinct elevation, often with a rounded profile, 1 km or more in diameter that is the geomorphologic expression of a diapir formed by vertical intrusion of salt. Commonly found in a PROVINCE of similar features.

SAND RIDGE*

An elongated feature of unconsolidated sediment of limited vertical relief and sometimes crescent shaped. Commonly found in a PROVINCE of similar features.

SEA CHANNEL

An elongated, meandering depression, usually occurring on a gently sloping plain or FAN.

SEAMOUNT

A distinct generally equidimensional elevation greater than 1,000 m above the surrounding relief as measured from the deepest isobath that surrounds most of the feature.

SEAMOUNT CHAIN

A linear or arcuate alignment of discrete SEAMOUNTS.

SHELF

The flat or gently sloping region adjacent to a continent or around an island that extends from the low water line to a depth, generally about 200 m, where there is a marked increase in downward slope.

SHOAL*

A shallow elevation composed of unconsolidated material that may constitute a hazard to surface navigation.

SILL

A relatively shallow barrier between BASINS that may inhibit water movement.

盐丘*

外轮廓常呈圆形，直径在1 km及以上边界清晰的海底高地。盐丘属地貌学术语，由垂直侵入的盐岩体构成。这类地理实体通常集中分布，构成一个**特征区**。

沙脊*

由有限垂直起伏的未固结沉积物构成的狭长地理实体，有时呈新月形。这类地理实体通常集中分布，构成一个**特征区**。

海底水道

狭长且蜿蜒曲折的海底洼地，通常发育在坡度平缓的海底平原或**海扇**上。

海山

轮廓清晰可辨、形状大致规则的海底高地，从环绕其主体的最大等深线起算，至顶部最大高差大于1 000 m。

海山链

呈线状或弧线状排列的离散**海山群**。

陆架

毗邻大陆或围绕岛屿周边，地势平坦或坡度平缓的海底区域，其范围从低潮线开始直到坡度显著增加处，一般延伸至水深约200 m附近。

浅滩*

由未固结物质构成的浅水高地，可能会危及海面航行安全。

海槛

相邻**海盆**之间水深相对较浅的隔挡地形，往往阻碍水体运移。

SLOPE

The sloping region that deepens from a SHELF to the point where there is a general decrease in gradient.

SPUR

A subordinate RIDGE protruding from a larger feature.

TERRACE

A flat or gently sloping region, generally long and narrow, bounded along one edge by a steeper descending slope and along the other by a steeper ascending slope.

TRENCH*

A long, deep, asymmetrical depression with relatively steep sides, that is associated with subduction.

TROUGH

A long depression generally wide and flat bottomed with symmetrical and parallel sides.

VALLEY

An elongated depression that generally widens and deepens down-slope.

陆坡

水深不断加大的海底斜坡区，其范围从**陆架**外缘开始一直延伸到坡度平缓的深海洋底为止。

山嘴

从较大地理实体中延伸出来的从属**海脊**。

阶地

地形平坦或坡度平缓的狭长海底区域，通常一侧以陡峭的下降斜坡为界，另一侧则以陡峭的上升斜坡为界。

海沟*

长而深邃的非对称海底洼地，两侧边坡较陡，成因与板块俯冲作用有关。

海槽

长而宽阔且底部较平坦的海底洼地，两侧边坡对称且平行。

海谷

地形顺坡而下，宽度随深度不断加大的伸长海底洼地。

II. GENERIC TERMS USED FOR HARMONIZATION WITH OTHER GAZETTEERS

Notes: the generic terms in this section are used for some features in the GEBCO Gazetteer and/or in other gazetteers. They are kept in this publication to facilitate harmonization between gazetteers. However, they are considered obsolete and their use is not recommended for new feature names.

ABYSSAL HILL

An isolated small elevation on the deep seafloor.

ARCHIPELAGIC APRON

A gentle SLOPE with a generally smooth surface of the seafloor, characteristically found around groups of islands or SEAMOUNTS.

BORDERLAND

A region adjacent to a continent, normally occupied by or bordering a SHELF and sometimes emerging as islands, that is irregular or blocky in plan or profile, with depths well in excess of those typical of a SHELF.

CAP

(See BANK)

CHANNEL

(See SEA CHANNEL)

CONE

(See FAN)

II. 与其他词典兼容的通名

说明： 本节中列出的通名用于某些在全球通用大洋水深图计划（GEBCO）海底地名词典和/或其他地名词典中收录的海底地名，但目前已停止使用。将其保留在本出版物中以便于不同地名词典之间的兼容，但不建议再将其应用于新的海底地理实体命名。

深海丘

位于深海洋底的小型孤立高地。

群岛裙

坡度平缓、底部地形平坦的**陆坡**，通常分布在群岛或**海山群**周围。

边缘地

毗邻陆地的区域，常被陆架占据或紧邻陆架，有时以群岛形式露出水面，平面上或外轮廓呈不规则状或块状，而水深则远超过典型**陆架**的水深。

海角

（见**滩**）

水道

（见**海底水道**）

海锥

（见**海扇**）

CONTINENTAL MARGIN

The zone, generally consisting of SHELF, SLOPE and CONTINENTAL RISE, separating the continent from the deep seafloor or ABYSSAL PLAIN or PLAIN. Occasionally a TRENCH may be present in place of a CONTINENTAL RISE.

CONTINENTAL RISE

A gently sloping region that extends from oceanic depths to the foot of a continental SLOPE.

CONTINENTAL SHELF

(See SHELF)

CONTINENTAL SLOPE

(See SLOPE)

DISCORDANCE

An area of the seafloor within a MID-OCEANIC RIDGE with rough and disordered morphology.

FRACTURE ZONE SYSTEM*

A group of closely spaced FRACTURE ZONES, which can also be called FRACTURE ZONE PROVINCE.

GROUND

(See BANK)

MEDIAN VALLEY

The axial depression of the MID-OCEANIC RIDGE.

MID-OCEANIC RIDGE

The linked major mid-oceanic mountain systems of global extent.

大陆边缘

通常由**陆架**、**陆坡**和**大陆隆**组成，将大陆和深海洋底或**深海平原**、**平原**分隔开来，**大陆隆**偶尔会被**海沟**所代替。

大陆隆

从**大陆坡**坡脚延伸至深海洋底的平缓斜坡区域。

大陆架

（见**陆架**）

大陆坡

（见**陆坡**）

不整合面

洋中脊内部带有粗糙无序地貌的海底区域。

断裂带系*

由海底一组相间分布但相对紧密的**断裂带**组成的地理实体群体，也称**断裂带特征区**。

底

（见**滩**）

中央裂谷

沿**洋中脊**轴部发育的海底洼地。

洋中脊

位于全球大洋中央、连绵不断的山脉系统。

PASS

(See SADDLE)

PASSAGE

(See GAP)

PLAIN

An extensive, flat or gently sloping region, usually found at depths less than 4,000 m.

PROMONTORY

A major SPUR-like protrusion of the CONTINENTAL SLOPE extending to the deep seafloor. Characteristically, the crest deepens seaward.

RE-ENTRANT

A prominent indentation in a SHELF-EDGE.

SCARP

(See ESCARPMENT)

SEA VALLEY

(See VALLEY)

SEABIGHT

(See VALLEY)

SEACHANNEL

(See SEA CHANNEL)

SEAMOUNT GROUP

Several closely spaced SEAMOUNTS not in a line, which can also be called SEAMOUNT PROVINCE.

隘口

（见**鞍部**）

山口

（见**裂谷**）

海底平原

地形平坦或坡度平缓的开阔海底，水深通常小于4 000 m。

岬

从**大陆坡**突出，一直延伸到深海洋底、以山嘴状为主的地貌单元，顶部水深向深海方向加大。

凹角

陆架外缘上的明显凹口。

陡崖

（见**海底崖**）

海谷

（见**海谷**）

海湾

（见**海谷**）

海底水道

（见**海底水道**）

海山组

由几个相邻且不成线状排列的**海山**组成，也称**海山特征区**。

SHELF BREAK

(See SHELF-EDGE)

SHELF-EDGE

The line along which there is a marked increase in slope at the seaward margin of a SHELF. Also called SHELF BREAK.

SUBMARINE VALLEY

(See VALLEY)

TABLEMOUNT

(See GUYOT)

陆架坡折

（见**陆架外缘**）

陆架外缘

陆架外部边界带，沿此边界**大陆架**向海一侧坡度明显增大。也叫**陆架坡折**。

海底谷

（见**海谷**）

桌状海山

（见**平顶海山**）

Chinese Alphabetical Index of the Chinese Terms Shown in the foregoing List of "UNDERSEA FEATURE TERMS AND DEFINITIONS", with Cross-References to the English Terms.

"海底地理实体通名及其定义"中按汉语拼音字母排序的通名索引表—英文/中文版

I. GENERIC TERMS I. 通名

MOUND	矮丘
SADDLE	鞍部
APRON	冲积裙
FRACTURE ZONE	断裂带
TROUGH	海槽
LEVEE	海堤
PEAK	海底峰
PLATEAU	海底高原
SEA CHANNEL	海底水道
CANYON	海底峡谷
ESCARPMENT	海底崖
TRENCH	海沟
VALLEY	海谷
RIDGE	海脊
SILL	海槛
RISE	海隆
BASIN	海盆
HILL	海丘
SEAMOUNT	海山
SEAMOUNT CHAIN	海山链
FAN	海扇
HOLE	海穴
DEEP	海渊
MOAT	环形洼地
PINNACLE	尖礁
REEF	礁
TERRACE	阶地

GAP	裂谷
SHELF	陆架
SLOPE	陆坡
MUD VOLCANO	泥火山
GUYOT	平顶海山
SHOAL	浅滩
SAND RIDGE	沙脊
SPUR	山嘴
ABYSSAL PLAIN	深海平原
CALDERA	塌陷火山口
BANK	滩
PROVINCE	特征区
SALT DOME	盐丘
KNOLL	圆丘

II. GENERIC TERMS USED FOR HARMONIZATION WITH OTHER GAZETTEERS

II. 与其他词典兼容的通名

PASS	隘口
RE-ENTRANT	凹角
BORDERLAND	边缘地
DISCORDANCE	不整合面
CONTINENTAL MARGIN	大陆边缘
CONTINENTAL SHELF	大陆架
CONTINENTAL RISE	大陆隆
CONTINENTAL SLOPE	大陆坡
GROUND	底
SCARP	陡崖
FRACTURE ZONE SYSTEM	断裂带系
SUBMARINE VALLEY	海底谷
PLAIN	海底平原
SEACHANNEL	海底水道
SEA VALLEY	海谷

CONE	海锥
PROMONTORY	岬
CAP	海角
SEAMOUNT GROUP	海山组
SEABIGHT	海湾
SHELF BREAK	陆架坡折
SHELF-EDGE	陆架外缘
ARCHIPELAGIC APRON	群岛裙
PASSAGE	山口
ABYSSAL HILL	深海丘
CHANNEL	水道
MID-OCEANIC RIDGE	洋中脊
MEDIAN VALLEY	中央裂谷
TABLEMOUNT	桌状海山

Appendix A

User's Guide for Preparation of Undersea Feature Name Proposals to the GEBCO Sub-Committee on Undersea Feature Names (SCUFN)

1. INTRODUCTION

The preparation of undersea feature name proposals should follow the guidelines contained in this publication (B-6). An Undersea Feature Name Proposal Form should be completed in English in accordance with the requirements specified in this publication and forwarded, preferably in a digital form, to IHO or IOC, no later than two months before an annual SCUFN meeting, in order to be considered by SCUFN members in advance of the meeting. SCUFN meeting dates can be found at www.iho.int＞IHO Council, Committees & WGs＞SCUFN.

2. PROCEDURE

2.1 Proposal Selection

- Identify unnamed features: first identify the position, extent and morphology of the feature and then certify that the selected feature has not already been named in the GEBCO Gazetteer (see www.gebco.net).

- Identify supporting data: single and/or multibeam bathymetric data, geophysical data, present and historical nautical charts, and other acquired data which can reflect the morphology of the undersea feature. This information should be based on reliable source data.

- Identify the metadata: check and verify the metadata information regarding the supporting data, including the survey dates, name or program, vessels, entities or persons involved, type and accuracy of the instruments, and so on.

附录 A

全球通用大洋水深制图计划（GEBCO）海底地名分委会（SCUFN）海底地名提案编制用户指南

1. 引言

海底地名提案应遵循本出版物（B-6）中规定的指南进行编制。提案表格应按照本出版物中的具体要求采用英文填写，最好以电子版形式提交至国际海道测量组织（IHO）或政府间海洋学委员会（IOC）。为了使海底地名分委会（SCUFN）委员能够在会前有充分时间对提案进行预审，提案提交时间不应晚于当年海底地名分委会（SCUFN）会议召开日期的前2个月。海底地名分委会（SCUFN）会议召开日期可在国际海道测量组织（IHO）官网下的海底地名分委会（SCUFN）主页中查询（www.iho.int＞IHO Council, Committees & WGs＞SCUFN）。

2. 编制程序

2.1 提案选择

- 识别尚未命名的地理实体：首先明确该地理实体的位置、范围和形态，然后确认所选定的地理实体在GEBCO海底地名词典中尚未被命名（见 www.gebco.net）。

- 确定支持数据：包括单波束和/或多波束水深地形数据、地球物理数据、现今和历史海图资料，以及其他能够反映该地理实体形态特征的数据资料。数据来源需真实可靠。

- 确定元数据：检查并验证提案支持数据的相关元数据信息，包括调查时间、调查项目或名称、调查船舶、参与调查的机构或人员、调查仪器的类型和精度等。

Note: Proposers are encouraged to release their bathymetric data, along with the associated metadata, to the IHO Data Centre for Digital Bathymetry (DCDB—see www.ngdc.noaa.gov/iho/).

2.2 Completing the Undersea Feature Name Proposal Form

- **Names Proposed:** composed of specific and generic terms. The specific terms are chosen by the proposer according to the relevant provisions in item Ⅱ, "Principles for Naming Features", sub-item A "Specific terms". The generic terms reflect the physiography of the feature and they should be selected from the list "Undersea Feature Terms and Definitions", sub-item Ⅰ "Generic Terms".

- **Geometry that best defines the feature:** geometry will be used to display and describe the undersea feature in the GEBCO Gazetteer. It should be a point, line, polygon, multiple points, multiple lines, multiple polygons or a combination of geometries. A primary geometry is assigned to a given generic term and, when appropriate, a secondary and tertiary geometry. See details in "SCUFN Generic terms—List of Allowed Geometries" (see www.iho.int/mtg_docs/com_wg/SCUFN/SCUFN_Misc/Feature_Geometries.xls). The coordinate of a feature whose geometry is limited to a point should be located in the centre or at the summit of the feature; for a feature whose geometry is a line, the coordinates should reflect the trend of the feature and; for a feature whose geometry is a polygon, the coordinate points should show the outline of the feature and the last coordinate point must be the same as the first one.

- **Coordinates:** geographic coordinates in Latitude S/N and Longitude E/W (degree, minute and decimal minute), Datum: WGS84. Example: Lat. 34°37.80′ S-Long. 028°52.17′ W. It is recommended that coordinates are expressed preferably in degrees, minutes and decimal minutes and

2.2 海底地名提案表填写

- **拟命名：** 由专名和通名组成。专名由提案人依据本文第Ⅱ款"海底地名命名原则"中的A分条款"专名"相关原则采词。通名反映地理实体的自然地理特征，应从本文"海底地理实体通名及其定义"中的Ⅰ分条款"通名"中选择。

- **最佳界定拟命名海底地理实体的几何图形：** 收录于GEBCO海底地名词典中的地理实体，需要以几何图形来展示和描绘，可为点、线、多边形、多点、多线、多个多边形或多个几何形状的组合。就给定通名，通常赋予一级几何图形，合适时也可以赋予二级和三级几何图形。具体参考"海底地名分委会（SCUFN）通名——允许使用的几何图形清单"（见：www.iho.int/mtg_docs/com_wg/SCUFN/SCUFN_Misc/Feature_Geometries.xls）。对于几何图形为点状的地理实体，该点坐标应位于地理实体中心或顶点；几何图形为线状的地理实体，坐标选取应能反映地理实体的走向趋势；几何图形为多边形的地理实体，多边形拐点坐标选取应能反映地理实体的轮廓，且坐标点要首尾相连。

- **坐标：** 地理坐标用纬度S/N和经度E/W表示（以度、分和分的小数表示），WGS-84坐标系。例如：纬度34°37.80′ S；经度28°52.17′ W。建议坐标最好以度、分和分的小数表示，并附矢量文件。在这种情况

accompanied with shape files. In that case, the number of vertices per line or polygon should be limited to 50, which is sufficient to depict the concerned feature. Further, the number of decimals should not exceed five, thus providing enough resolution to locate this fictitious line.

- **Feature Description:** specify maximum and minimum water depths over the feature, which should be extracted from a trackline sounding or a bathymetric terrain model derived from in situ soundings rather than from a predicted bathymetric grid developed with satellite altimetry data; the total relief, which is the difference between the maximum and minimum depths; the steepness which is the ratio of the vertical height and the horizontal distance, expressed in degrees; the shape as round, square, triangle, elliptical, or U/V in the case of a canyon; and the dimensions of the feature specifying its length and width. The units of size and depths should be kilometers (km) and meters (m), respectively.

- **Associated Features:** List of names of existing recognized features specifically in the GEBCO Gazetteer which are in close and medium proximity or associated with the proposed feature are to be provided. See 3.1 (index maps).

- **Chart/Map References:** the number of a map or nautical chart where the proposed feature is shown and named, or only shown, should be identified in this item. If not shown or named on any existing chart or map, the number of an INTernational (INT) and/or national chart in which the feature falls, may be indicated (see Catalogue of INT charts: htte://iho.int/iho _pubs/IHO_Download.htm#S-11).

- **Reason for Choice of Name:** this item must contain a detailed description as to the reason for having chosen the specific term, following the rules which appear in item Ⅱ, sub-item A. Historical information regarding the origin of the chosen name should be provided. Names

下，对于几何图形为线或多边形的地理实体，每条线或多边形拐点坐标数应不超过50个，这足可以描述地理实体的形状。另外，小数位数不要超过5位，因为这足以保证轮廓线的定位精度。

- **拟命名海底地理实体描述：** 标明地理实体的最大和最小水深值，该值应从走航式测深数据或实测水深数据建立的水深地形模型中提取，不应从由卫星测高资料推算的水深网格数据中提取；总起伏，是指最大水深和最小水深的差值；坡度，指地理实体坡面的垂直高度和水平距离的比值，用度表示；形状，如圆形、方形、三角形、椭圆形，或描述海底峡谷的U/V形；尺度/大小范围，指海底地理实体的长度和宽度，大小和深度分别用千米（km）和米（m）表示。

- **相关地理实体：** 提供已在GEBCO海底地名词典中收录，且与拟命名的海底地理实体邻近或存在地形关联的地名列表。见3.1（索引图）。

- **参考海图/地图：** 标识显示拟命名地理实体的地图或海图编号。如果没有显示或命名该地理实体的海图或地图，可填写包含该地理实体所在位置的国际通用海图（INT）和/或国家公开出版的海图图幅号。（国际通用海图编目查询网址：htte://iho.int/iho_pubs/IHO_Download.htm#S-11）。

- **选择地名的理由：** 详细描述专名采词的理由。专名的选择应遵循"海底地名命名标准指南"第Ⅱ款A分条款中"专名"的相关规定。应提供与专名选择依据有关的历史信息。专名选择应优先考虑与其相关

should preferably be associated with a geographical feature. When a ship name is proposed, it should preferably be the name of the discovering ship or the one that surveyed and verified the feature. In the case of a name proposed after a living person, that person should have made a recognized outstanding or fundamental contribution to ocean sciences; accordingly, his/her biography should be attached.

- **Discovery Facts:** the discovery date and discoverer ship or individual, if known.

- **Supporting Survey Data, including Track Controls:** information regarding the survey and data. Date(s) of survey(s); survey ship / platform; sounding equipment (brand and model of the singlebeam or multibeam or both); positioning system (GNSS, etc.); estimated horizontal accuracy; survey trackline spacing.

- **Proposer(s):** name of the proposer(s) or the institution who prepared and submitted the feature name; date of forwarded proposal, E-mail, organization and address.

- **Remarks:** any other information considered important and supporting information such as maps, bathymetric grids, 3D models, charts, scientific publications, information on pre-existing published name(s) for the feature-if known-and so on. When a generic term with genetic implications is proposed, geological and/or geophysical evidence as well as bathymetric data must be provided.

3. SUPPORTING MAPS

Additional background documents should be provided in order to better support the proposal submitted to SCUFN. Maps with specific information should be included in the proposal as in the examples given below.

Note: All graphics shown as examples are based on multibeam bathymetric data. However name proposals can be submitted to SCUFN, which are based on single beam bathymetry only, as long as there is sufficient data coverage.

的地理实体名称。如果采用船名命名时，该船应是首先发现该地理实体的船只，或者是调查并核实该地理实体的船只。如果采用在世人名命名，此人必须对海洋科学做出过杰出贡献，并附其个人履历。

- **发现事实**：发现提案地理实体的时间和调查船或发现者，如已知。

- **支持获得本次发现的调查资料，包括测线控制**：涉及地形调查和数据获取的相关信息，包括调查日期、调查船或平台、测深设备（单波束和/或多波束设备的类型及型号）、定位系统（GNSS等）、估计水平精度、测线间隔。

- **提案人**：编制并提交该提案的人员或机构名称、提交日期、电子邮箱、单位和地址。

- **备注**：其他任何重要且对提案起支持作用的信息，如地图、水深网格数据、三维模型、海图、学术出版物、已知该地理实体曾在公开出版物中命名的相关资料。若提案中地理实体的通名具有成因含义，除了提供水深数据外，还须提供地质和/或地球物理方面的佐证资料。

3. 支撑图件

为了更好地支撑提交至海底地名分委会（SCUFN）的海底地名提案，提案人应在提交提案表格的同时提供更多的背景资料，包括标识具体信息的支撑图件，示例如下。

说明：所有示例图件均基于多波束测深数据生成。然而，提交至海底地名分委会（SCUFN）的海底地名提案中的支撑图件也可以只基于单波束数据编绘，但必须保证数据能够充分覆盖。

3.1 Small-scale index map showing the location of the proposed feature on a regional scale.

3.1 小比例尺索引图，在区域尺度展示提案地理实体的所在位置。

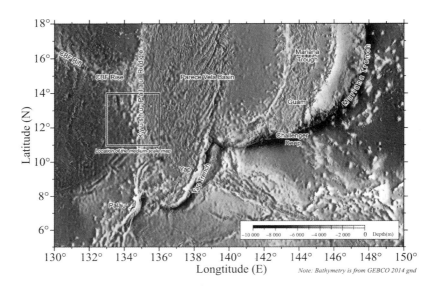

3.2 Medium-scale map, as considered appropriate, that help SCUFN understand the general tectonic and morphological context of the proposed feature. The map should show any internationally-recognized features, and/or any existing undersea feature names.

3.2 适当范围的中比例尺图，可帮助海底地名分委会（SCUFN）了解提案地理实体周围的地质构造和地貌形态。图件应包含制图范围内国际公认的地理实体和/或已有的海底地名。

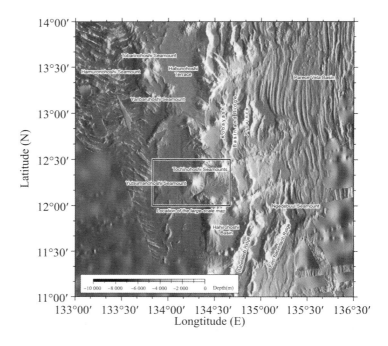

| 3.3 Large-scale track line map showing all existing information in the feature proposal area. | 3.3 大比例尺航迹图，显示提案地理实体范围内的所有测量航迹。 |

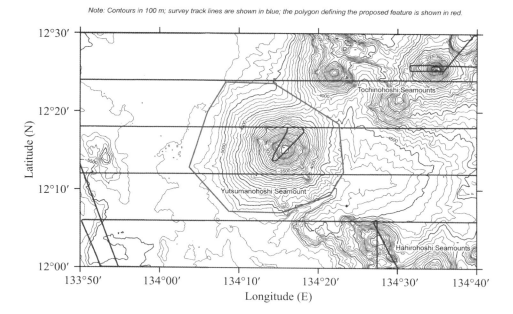

Note: Contours in 100 m; survey track lines are shown in blue; the polygon defining the proposed feature is shown in red.

| 3.4 Large-scale bathymetric map showing depth contours specifying the interval contour value, or a bathymetric shaded image with a depth colour legend, or both. | 3.4 大比例尺水深图，即展示按一定等深值间隔标注深度的等深线图，或带有深度彩色渐变图例的渲染水深图，或两者兼备。 |

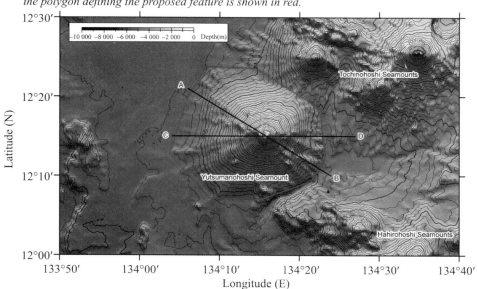

Note: Contours in 100 m; location of two profiles A-B and C-D is shown in black solid line; the polygon defining the proposed feature is shown in red.

3.5	2D bathymetric oriented profile (s) of the proposed feature. The location of the profile (s) should be indicated on the large-scale bathymetric map shown above (See 3.4).	3.5 基于水深的二维剖面图，该剖面位置应标注在上述大比例尺水深图上（见3.4）。

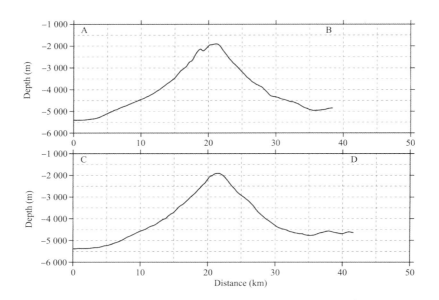

3.6	3D bathymetric image that best displays the entire picture of the proposed feature.	3.6 能够最好展示拟命名地理实体全貌的三维水深图。

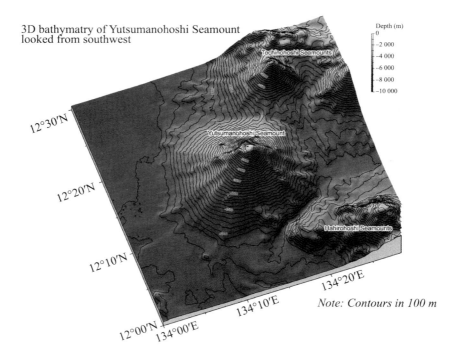

4. CONCLUSION

The undersea feature name proposal form should be completed with all available and reliable information in order to better define the submarine feature. As the number of undersea feature name proposals submitted to SCUFN has been increasing over the years, the more complete the proposal, the more consistent and rapid will be the response of SCUFN, thus avoiding having to make additional requests to the proposer. Once the proposal is approved, the feature name will be inserted in the "GEBCO Gazetteer of Undersea Feature Names".

4. 结论

为了更好地界定拟命名的海底地理实体，海底地名提案表应填写所有可用和可靠的信息。鉴于近年来提交至海底地名分委会（SCUFN）的提案数量逐年增加，提案信息越完整，海底地名分委会（SCUFN）就越能做出一致且快速的回应，从而避免向提案人提出额外的信息补充要求。提案一经审议通过，该海底地名就会被收录至"GEBCO 海底地名词典"中。

Appendix B

Procedure for the Adoption of Undersea Feature Names Proposed by National Geographical Naming Authorities
(Fast-Track Procedure)

1. SCOPE

1.1 This procedure applies for undersea feature names proposed by national geographical naming authorities that are recognized by SCUFN[①].

1.2 This procedure applies for undersea feature names that are in long term (25 years or longer) common use and appear on published charts, maps or in scientific literature.

2. PRELIMINARY CONDITIONS

2.1 SCUFN will maintain a list of recognised national geographical naming authorities, with references/links to the appropriate national regulations and authority under which they function.

2.2 Any national geographical naming authority wishing to be recognized and listed in the SCUFN register shall provide SCUFN, via the Secretary, with references/links to the appropriate national regulations and authority under which it functions together with its rules of procedure or guidelines for naming undersea feature names.

2.3 Applications for recognition by SCUFN as a national geographical naming authority for the purposes of this procedure, can be made at any time and will be considered at each meeting of SCUFN.

2.4 In accordance with the Rules of Procedures for SCUFN, proposals that are politically sensitive will not be considered.

[①] A national Hydrographic Office can also play this role.

3. METHODOLOGY

Submission by SCUFN-recognized national geographical naming authority

Step 1: Recognised national geographical naming authority proposes undersea feature name (or a set of names) as adopted under their governing rules to SCUFN Secretary, including:

3.1 a basic list of proposed names with their coordinates and shapefiles;

3.2 the location of the features to be named (graphics, maps, chartlets that clearly depict the feature, its name and bathymetric data coverage are required if the feature cannot be defined by the GEBCO grid bathymetry);

3.3 confirmation that the names have been designated under the national geographical naming authority's rules of procedure or guidelines for naming undersea features.

Submissions may be made at any time.

Maximum: no more than 50 names per year.

Step 2: SCUFN Secretary confirms eligibility of submitting authority and circulates proposals to SCUFN members for validation review. The review will consider whether each name conforms to SCUFN (B-6) guidelines, taking into consideration the list of generic terms that can be used for harmonizing gazetteers.

When proposals are accepted by SCUFN Members, the undersea feature names can be included directly in the GEBCO Gazetteer without further review or consideration by SCUFN "fast-track" route option.

Step 3: If no objection is raised by any SCUFN member within two months, then Secretary will include the submitted name(s) of the feature(s) directly in the GEBCO Gazetteer and provide

3. 方法

提案需由海底地名分委会（SCUFN）认可的国家地名管理机构提交。

第一步：海底地名分委会（SCUFN）认可的国家地名管理机构向海底地名分委会（SCUFN）秘书提交依据该国管理规则通过的单个（或一组）海底地名提案，具体包括：

3.1 一份附有提案地理实体的坐标和矢量文件的基本清单；

3.2 拟命名海底地理实体的位置信息（能清晰描述地理实体形态特征的图形、地图、图表等，如果该地理实体在GEBCO水深图中无法清晰辨认，则需要提供覆盖地理实体范围的实测水深数据）；

3.3 确认提案是由国家地名管理机构根据其海底地名有关议事规则或指南编制的。

可随时提交海底地名提案。

每年提交的提案数量不超过50个。

第二步：海底地名分委会（SCUFN）秘书在确认海底地名提案提交机构的资格后，将提案分发至海底地名分委会（SCUFN）委员们进行有效性审议。主要审议内容包括命名是否符合海底地名分委会（SCUFN）（B-6文件）指南，是否考虑了与其他地名词典中的通名保持一致。

提案一旦被海底地名分委会（SCUFN）委员们接受，该海底地名则不需再经海底地名分委会（SCUFN）进一步审议，将直接被纳入GEBCO海底地名词典中"快速审议"程序。

第三步：如果所有海底地名分委会（SCUFN）委员在两个月内均未提出异议，海底地名分委会（SCUFN）秘书处将直接把提交的海底地名提案收录至GEBCO海底地名词典中，并

a summary report to SCUFN meetings on any proposals accepted or rejected under this procedure during the intersessional period "fast-track" route option.

Step 4: If any objection is raised on any name or feature by a SCUFN member, the Secretary will notify all SCUFN members and invite an ad hoc review panel to re-consider and attempt to resolve any objections raised to the satisfaction of the submitting organization and any objecting Member of SCUFN.

Step 5: If the objection is resolved, the rapporteur of the ad hoc review panel shall report to the next SCUFN meeting and the Secretary will include the agreed feature and name in the GEBCO Gazetteer; or

If the objection cannot be resolved, the rapporteur of the ad hoc review panel shall report to the next SCUFN meeting "normal" route option. The Leader shall provide a briefing and present any recommendations on whether the matter should be considered further by the SCUFN, or dismissed. As a matter of principle, the plenary SCUFN meeting should normally follow the recommendations of the ad hoc review panel. The plenary SCUFN meeting will make a final decision on the recommendations.

4. AD HOC REVIEW PANELS

4.1 Ad hoc review panels shall comprise three or more members of SCUFN, on a voluntary basis. The members, including the rapporteur, shall be designated by the Secretary, in consultation with the Chair of SCUFN if necessary.

4.2 As all SCUFN members represent their parent organization (IHO, IOC), and not any national naming authority, the composition of the panel will be decided on a case-by-case basis for efficiency purposes.

在当年的海底地名分委会（SCUFN）会议总结报告中说明在闭会期间提案采纳和拒绝情况。

第四步：如有海底地名分委会（SCUFN）委员对提案提出的地名或地理实体本身提出任何异议，海底地名分委会（SCUFN）秘书会将此情况通知所有海底地名分委会（SCUFN）委员，并成立特设审查小组进行重新审议，力争获得使提案提交机构和提出反对意见的海底地名分委会（SCUFN）委员均满意的解决方案。

第五步：若异议得到解决，特设审查小组报告员将向下一次海底地名分委会（SCUFN）会议报告，同时秘书将已经同意的海底地理实体和地名收录至GEBCO海底地名词典中。或

若异议未得到解决，特设审查小组报告员将向下一次海底地名分委会（SCUFN）会议报告相关情况"正常审议"程序。小组负责人需作简要说明并就海底地名分委会（SCUFN）是否应进一步讨论或驳回该提案提出建议。原则上，海底地名分委会（SCUFN）全体会议通常会采纳特设审查小组的建议，并依据建议作出最终决定。

4. 特设审查小组

4.1 特设审查小组由三个或以上海底地名分委会（SCUFN）委员在自愿基础上组成。小组成员中的报告员由秘书指定，必要时需与海底地名分委会（SCUFN）主席协商。

4.2 由于所有的海底地名分委会（SCUFN）委员仅代表他们各自所属国际组织（IHO、IOC），而不是任何国家地名管理机构，为了提高效率，特设审议小组成员的组成将根据具体情况确定。

4.3 The Secretary shall provide the members of an ad hoc review panel with all relevant information in order for them to undertake their work, including, as appropriate:

Specific Name:
Generic term:
Latitude:
Longitude:
References:
History: when first appeared on charts and/or discoverered-where known.
Origin of name: notes on the reason for the name-where known.
Additional information: any relevant information such as chart/s maps and papers that are the key references for the name.
Max depth:
Min depth:
Total relief:
Dimensions:
Polygon / polyline: for GIS

It is essential that supporting bathymetric map(s)/chart(s)/diagram(s) are provided

4.3 为了便于开展工作，秘书将为特设审查小组的成员提供有关地名提案的所有相关资料，包括：

专名：
通名：
纬度：
经度：
参考材料：
命名历史：已知何时出现在海图上的和/或何时被发现的。
命名缘由：已知的命名理由记录。

其他信息：任何可作为地名提案关键参考的信息，如海图、地图、论文。

最大水深：
最小水深：
总起伏：
范围：
多边形/线：适用于地理信息系统软件的矢量文件。
水深图、海图、示意图等必要的支撑图件。

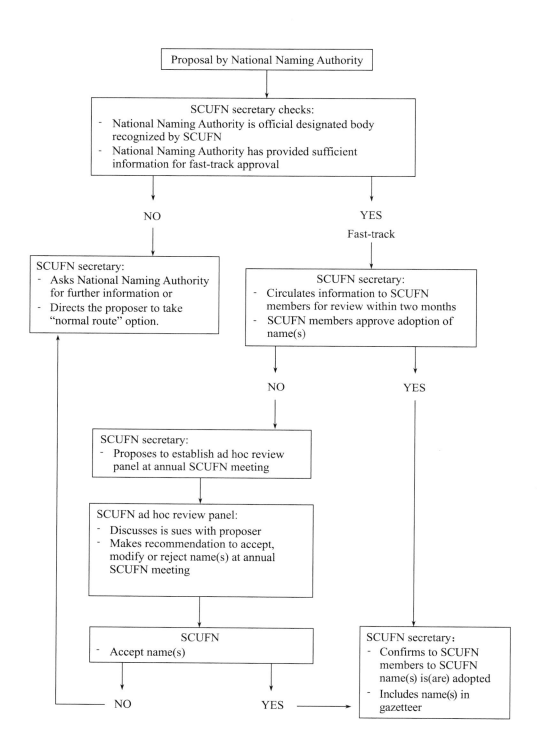

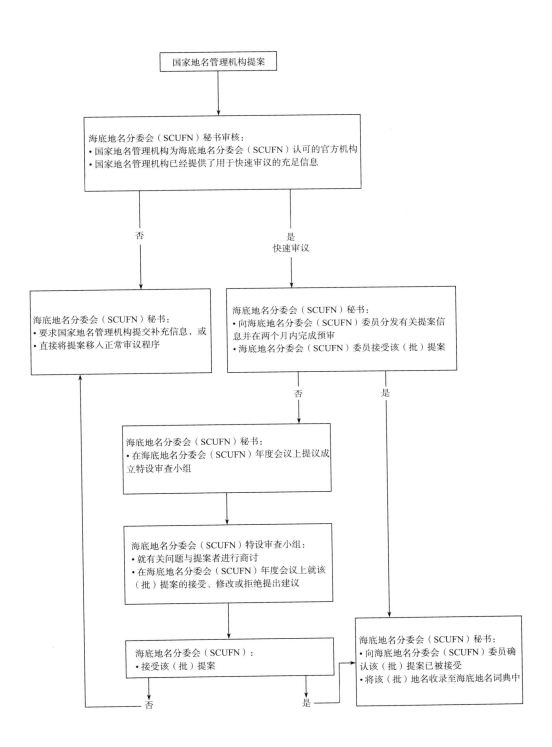